We need to eat foods in
the **dairy** group.

Dairy foods are made of milk.

Dairy foods give us **calcium**.

Dairy foods give us strong
bones.

Dairy foods give us healthy teeth.

We need two **servings** of
dairy foods each day.

We can drink milk.

We can eat yogurt.

We can eat cottage cheese.

We can eat Swiss cheese.

We can eat Cheddar
cheese.

We can eat mozzarella
cheese.

We can eat cream cheese.

We can eat ice cream.

Dairy foods keep me healthy.

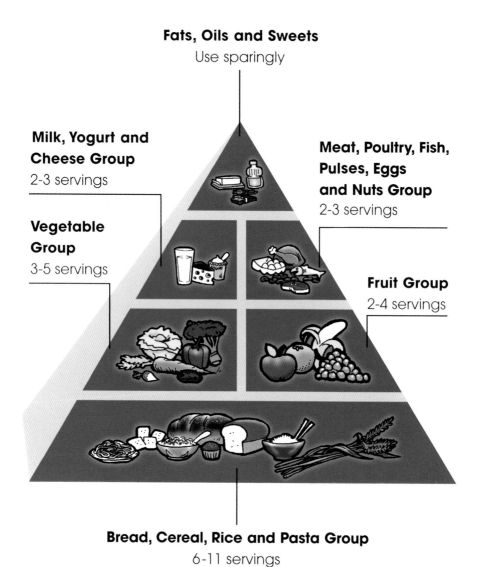

Fats, Oils and Sweets
Use sparingly

Milk, Yogurt and Cheese Group
2-3 servings

Meat, Poultry, Fish, Pulses, Eggs and Nuts Group
2-3 servings

Vegetable Group
3-5 servings

Fruit Group
2-4 servings

Bread, Cereal, Rice and Pasta Group
6-11 servings

Milk, Yogurt and Cheese Group

The food pyramid shows us how many servings of different foods we should eat every day. The milk, yogurt and cheese group is on the third level. The foods in this group are called dairy products. You need 2–3 servings of dairy products every day. You could drink a cup of milk or eat a slice of cheese. Dairy products are good for you because they have calcium. Calcium gives you strong bones and teeth.

Dairy Facts

The average British person drinks 85 litres of milk in a year.

There are many different kinds of milk. You could try whole milk, semi-skimmed milk, skimmed milk, buttermilk or chocolate milk.

There are 2.3 million dairy cows in Britain.

A cow must drink 8 litres of water to make one litre of milk.

The average cow makes 80 glasses of milk a day.

To get the same amount of calcium as you would from a 200ml glass of milk you would have to eat 11 portions of spinach or 15 portions of red kidney beans.

There are over 400 different kinds of cheese.

It takes 10 litres of milk to make 1 kilogram of cheese.

Glossary

 calcium – helps build and repair bones and teeth

 dairy – foods made from milk

 healthy – not sick; well

 servings – amounts of food

Index

This book was first published in the United States of America in 2003.

First published in the United Kingdom in 2008 by
Lerner Books,
Dalton House,
60 Windsor Avenue,
London SW19 2RR

Website address: www.lernerbooks.co.uk

This edition was updated and edited for UK publication by Discovery Books Ltd., Unit 3, 37 Watling Street, Leintwardine, Shropshire SY7 0LW

Words in **bold** type are explained in the glossary on page 22.

British Library Cataloguing in Publication Data

Nelson, Robin, 1971-
    Dairy. - (First step non-fiction. Food groups)
    1. Dairy products in human nutrition - Juvenile literature
    2. Dairy products - Juvenile literature
    I. Title
    641.3'7

    ISBN-13: 978 1 58013 387 6

The photographs in this book are reproduced through the courtesy of: © Todd Strand/Independent Picture Service, front cover, pp 2, 5, 8, 10, 11, 13, 14, 15, 16, 22 (top, bottom); © Milch & Markt, http://www.milch-markt.de, pp 3, 9, 12, 17; © Mitch Hrdlicka/PhotoDisc/Getty Images, pp 4, 22 (second from top); © Amos Morgan/PhotoDisc/Getty Images, p 6; © EyeWire/Getty Images, pp 7, 22 (second from bottom).

The illustration on page 18 is by Bill Hauser/Independent Picture Service.

Printed in China